The Adventures of
BIPLOB THE BUMBLEBEE

VOLUME 4

Biplob World Pvt. Ltd.

Publisher Ritika Talwar
Published by Biplob World Pvt. ltd.
© 2024 Biplob World Pvt. Ltd. All Rights Reserved.
Typeset in Sabon LT Std Roman

*Winner of NMCBI Best Childrens Author 2021

CONTENTS

Story 1: The Mystery Monster
  Animal Behaviour - Dogs

Story 2: Waste to Gold
  Origami - Recycling Paper

Story 3: The Championship
  Importance of 4R's: Refuse, Reduce, Reuse, Recycle

1
THE MYSTERY
MONSTER

"He has ruined everything! So many days of hard work and loving care, all destroyed in one afternoon!" wailed MS. Daisy.

Biplob too was shocked when he saw the flowerbeds MS. Daisy was crying over. Nearly all the plants were uprooted. The mud was churned up. In place of the plants, there were little stones and even a bone! "Who made this mess?" asked Biplob, very upset.

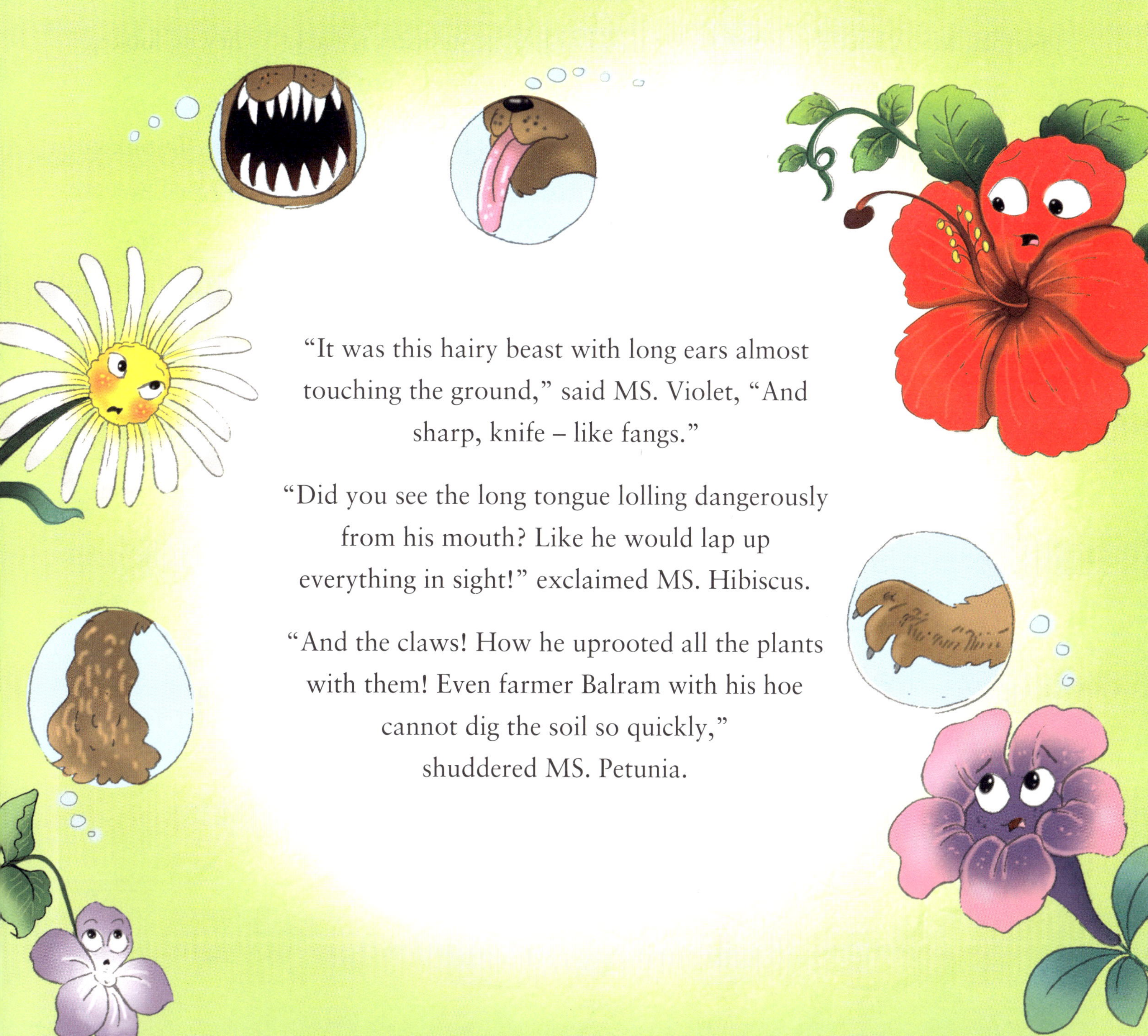

"It was this hairy beast with long ears almost touching the ground," said MS. Violet, "And sharp, knife – like fangs."

"Did you see the long tongue lolling dangerously from his mouth? Like he would lap up everything in sight!" exclaimed MS. Hibiscus.

"And the claws! How he uprooted all the plants with them! Even farmer Balram with his hoe cannot dig the soil so quickly," shuddered MS. Petunia.

Just then MS. Violet gasped, "Hush now. Look – the monster is back!" They all looked aghast as the dreaded monster came bounding into the garden.

Suddenly, Biplob let out a loud laugh. All the flowers looked at him, amazed. "I think he has gone mad with fear," whispered MS. Violet, "I have heard that happens when you are very scared."

"Oh I haven't lost my mind my innocent friends," said Biplob. "I call you innocent because you don't recognize a nice friendly dog when you see one!"

"Friendly dog?" asked the flowers, "what are you talking about?"
"Your monster is nothing but a friendly little Cocker Spaniel!" replied Biplob. The dog, which by now had reached the flowerbeds, started burying another bone in the ground.

Just then Biplob's old friends, Addy and Avantika, along with a friend, came running into the garden.

"Hi Biplob," said Addy, "have you seen our dog Floppy?
"Look, there he is, burying the bone which he stole from the lunch table," pointed Avantika, spotting Floppy's stubby tail.

"Hello there!" replied Biplob sadly, "Floppy has destroyed our flowerbeds in burying his bones. Look at all the plants he dug up."

Addy and Avantika were shocked to see the mess and apologised immediately. "What can we do to make this better?" they asked.

"Why don't you children get the spade and hoe and replant all the saplings that Floppy has uprooted?" suggested Biplob

"Yes, we'll do that if you show us how," replied Addy and Avantika eagerly, "Our friend Ved can help too."

They quickly collected the gardening tools from the shed and went to work.

Biplob and the flowers showed them how to dig out the stones and bones Floppy had buried in the soil and put the plants back in.

Soon, all the plants were neatly back in the flowerbed. The children added mulch and watered them for a speedy recovery.

"Well done!" beamed Biplob happily, "Now you just have to ensure that Floppy doesn't dig up the flowerbeds again to bury his bones."

Floppy guiltily put his head down. He knew he'd done something wrong.

"How do we stop Floppy from ruining the garden again?" asked Addy.

"Simple. We find him a nice spot where he can bury his things without disturbing the garden!" explained Biplob.

Biplob and the children found Floppy an
empty corner in the garden to
bury his bones.

Floppy happily danced around them,
wagging his tail. He was thanking them for
finding a safe spot for his treasure!

The children happily skipped back home
with Floppy as Biplob buzzed back to work
after solving yet another problem in his garden!

2
Waste to Gold

"Wait till the mayor sees my flower bed!" said
MS. Violet. "That
alone will win us the Purple Ribbon!"

"There she goes," giggled MS. Hibiscus, "Blowing her
own trumpet again."

"Yes," replied MS. Gladioli, "Like the rest of us do
nothing except preen in the sun all day!"

Just then, Avantika and Addy skipped into the garden.
"What are they arguing about?" asked Addy.

"They are bickering about the annual 'Best Garden Competition'. Every year, the mayor visits all the gardens in the district and felicitates the best garden with the Purple Ribbon," replied Biplob.

"Oh wow! That means your garden can win the competition," said an excited Avantika.

"Well, that entirely depends on what Jaggu has planned for this year," replied farmer Balram.

"Jaggu is the richest man in our village," explained Biplob to the puzzled children, "He wins every year because his garden is filled with expensive plants that he's bought from all over the world. Some plants are so exotic, we've not even heard their names, let alone seen them."

The children wore glum expressions. It seemed unfair that Jaggu would win simply because he was richer than everyone else!

"Only two days before school resumes," said Avantika. Aditya nodded glumly.
Just then, Biplob buzzed in, "Hey! What's this?!" he exclaimed, pointing excitedly at a table strewn with swans, frogs, airplanes, butterflies, dogs, even dinosaurs.
"Oh, that's Addy's origami," Avantika answered, bored. "Origami is a craft where paper is folded to form different figures.
Isn't it cool, Biplob?" asked Addy. "No, it's annoying! There's no place to keep my things. Everywhere, there's a paper animal or bird already perched!" Avantika complained. "Oh, really? What about your action figures?" Addy shot back.
Biplob ignored the squabbling kids. "Well, come along, dinner is ready," he said.

"Wake up guys!" said Biplob, buzzing into the children's room early next morning.

"Let us sleep! Last two days left of our vacation," mumbled Addy sleepily.

"No way Addy, I have a splendid plan to win the Purple Ribbon!" pirouetted Biplob.

"What plan?" Both Addy and Avantika sat up.

"Get ready and come for breakfast. I'll tell you there," replied Biplob, zooming out.

"What's your 'splendid plan' to win the Purple Ribbon?" asked farmer Balram over breakfast. "Last night I saw Addy's origami," replied Biplob, "It's so realistic, I thought the animals would jump off the table!" Addy blushed.

"How will that help us win the Purple Ribbon?" persisted farmer Balram. "We can decorate the garden with Addy's origami," explained Biplob. "That way, our garden will look interesting even without spending money like Jaggu!" "That's worth trying," replied farmer Balram, thoughtfully.

After breakfast Avantika set about collecting old newspapers for Addy to make the origami.

Meanwhile, Biplob and farmer Balram toured the garden to plan how the origami should be placed for maximum effect.

By evening Addy had made over 50 origami pieces!

Early next morning, with Biplob guiding them, they placed all the figures exactly as planned.

"My word! This looks beautiful!" exclaimed Ms. Daisy.

"Look at the cute dinosaurs in my patch!" gushed Ms. Hibiscus.

"And these birds add so much life to my flower bed," chimed in Ms. Gladioli.

"I guess these paper dogs are good too," added Ms. Violet, "If only they weren't made of old newspapers, though. Coloured paper would've been much nicer!" she sniffled.

The other flowers looked at each other knowingly and giggled.

The garden was decorated and ready for the mayor's visit.

All that Biplob, farmer Balram, Aditya, and Avantika could do now was wait for the mayor to arrive.

3
The Championship

The entire village had gathered in the square to welcome the Mayor.

The village headman announced, "There are two finalists from our village this year— farmer Jaggu, who won the Purple Ribbon last year, and farmer Balram."

"Ah, the two who competed last year too! I remember them well," replied the mayor, looking at them.

They all headed towards farmer Jaggu's garden first.

Jaggu's garden looked splendid with a sandalwood fence and large decorative gates with gold handles. It had exotic flowers like angel wing begonias, bromeliads, purple–blue marigolds and blue–ice calatheas.

"Welcome to my magnificent garden, Mr. Mayor!" boomed Jaggu, "You must be tired. Here, have some lemonade prepared from procida lemons imported from Italy and planted in my garden."

The mayor gratefully took a sip of the refreshing beverage.

"My flowers are a treat to the eye," boasted Jaggu, before glancing at farmer Balram and continuing, "Of course, some of us can learn gardening from the way they've been arranged."

"You've spent a small fortune here farmer Jaggu," remarked the mayor. Jaggu laughed, "Of course! For instance, every brick lining the flowerbeds is plated with gold."

"Truly looks magnificent. Let's move on to the next garden," replied the mayor.

On the way, Jaggu put an arm around farmer Balram and whispered, "Don't worry, old chap! Whenever I get new exotic plants, you can have my old plants for your garden. That way, you will at least come second. With me getting the Purple Ribbon, the first and second prize will always be won by our village!"

Farmer Balram looked very cross. Before he could reply, Biplob silenced him with a reassuring look.

"Well, what's this!" exclaimed the mayor, staring at the origami in farmer Balram's garden.

Jaggu replied before farmer Balram could react, "Please don't mind that these thingamajigs are made with old newspapers. Even though farmer Balram is not rich, he has a heart of gold. So what if he cannot afford gold coloured paper!"

"Those 'thingamajigs' are called origami," replied the mayor, "And please allow farmer Balram to talk about his garden."

"Thank you MR. Mayor, actually this is Biplob's idea.
He should explain it to you," farmer Balram said honestly.

"Mr. Mayor, all these origami figures have been created by our talented friend, Aditya. We strongly believe in reusing and recycling and so decorated the garden with origami created from old newspapers!" Biplob explained, as the flowers nodded.

The mayor went around the garden thoughtfully.

"Farmer Jaggu, your gorgeous garden has everything except two important elements — passion and love, which farmer Balram's garden has in abundance. This is what makes them think innovatively," said the mayor finally, "If this competition were just about money, we may as well give the Purple Ribbon to the richest farmer. Farmer Balram's plants may not be exotic but I have never seen happier looking flowers in my life."

"Farmer Balram's fence may have broken bricks, but they've been laid with his own hands," he added, "That is exactly what this competition is about!"
"So many have worked hard on this garden. Children, flowers and a bright bumblebee! Looking at their team effort, innovative thinking and passion, I am awarding this year's Purple Ribbon to farmer Balram!" declared the mayor, pinning the winner's ribbon on farmer Balram, who beamed with joy.

As Jaggu sulked away, farmer Balram offered the mayor some lemonade.

"Our lemonade is made from homegrown lemons which are as good as any other lemons in the world!" announced Biplob, as they all chuckled.

"Did you notice how the mayor looked only at my flower bed when he called our garden 'happy'?" boasted MS. Violet.

"Oh Violet!" exclaimed all the flowers. The sound of laughter once again filled the garden.

The Adventures of Biplob the Bumblebee : Book Series

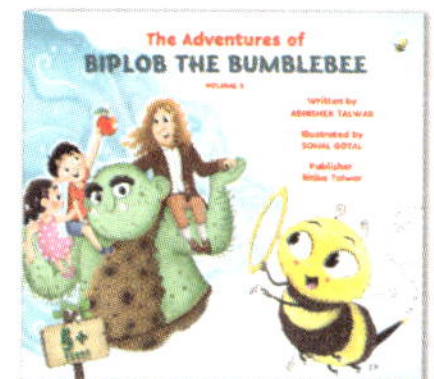

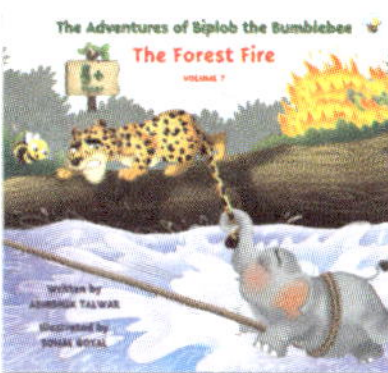

Biplob the Bumblebee : Early Learner Series

Detective Col. Zoro Series

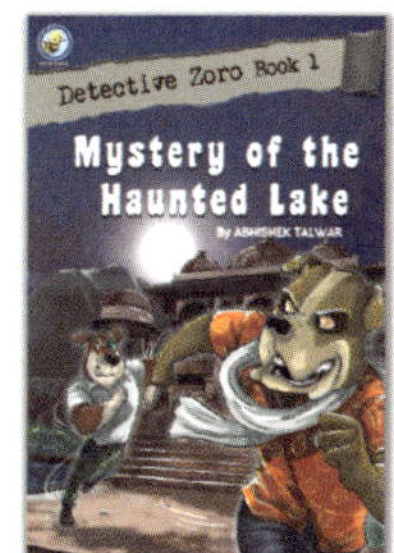

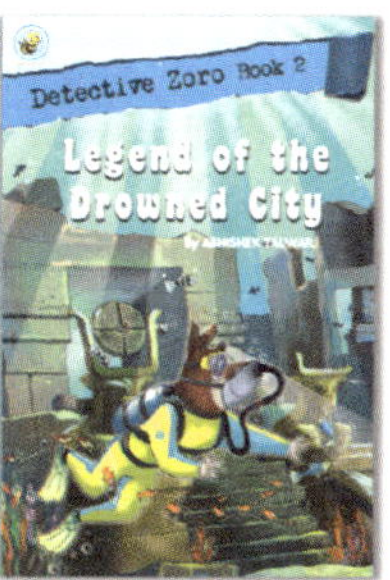

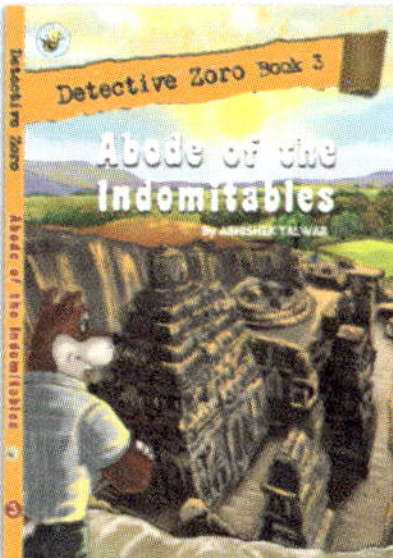

DRAW YOUR STORY

If you enjoyed reading about these adventures of Biplob the Bumblebee, then you will love what comes next! Read the paragraph below:

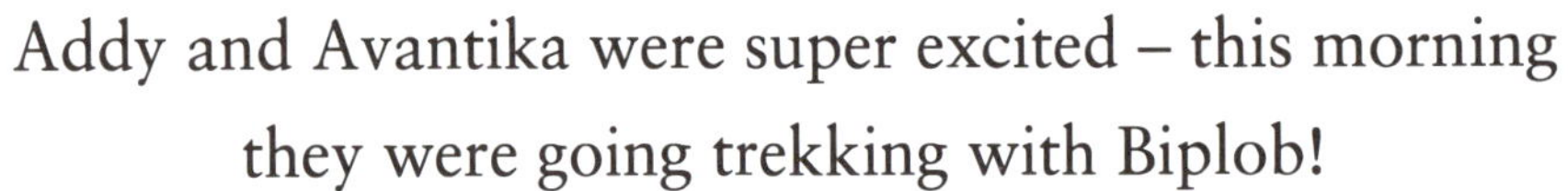

Addy and Avantika were super excited – this morning they were going trekking with Biplob!

They quickly finished breakfast and headed out with their water bottles and sandwiches. "Don't forget your caps," said Biplob as Floppy flapped his ears and barked excitedly.

They trooped out, passing through the village as it slowly woke up and came to life

Now, draw the scene described in it, on the opposite page. Remember, there is nothing right or wrong – use your imagination, draw and colour the scene, as you see it!

Scan this code to upload an image of your drawing
and you could get featured on our instagram page!

About the Author

Abhishek Talwar (AIEMA) is a certified environmentalist and author. He has the gift of narrating difficult concepts with simplicity and fun. This ease of narrative combined with the values of love for the environment, loyalty and friendship that shine through each story make these books truly entertaining and enriching for children.

"Dedicated to my loving wife, Ritika, for always believing.
And to my curious children, for inspiring these tales."

-Abhishek Talwar

About the Illustrator

Sonal Goyal completed her masters in fine arts from College of Art, Delhi. Having been creative head at a leading publishing house, she is now an independent artist. Her love for drawing and books has led her to creating cute and lovable illustrations that spread smiles.

www.biplobworld.com